Optimum Egg Quality
A Practical Approach

Optimum Egg Quality
A Practical Approach

Jeffrey A. Coutts
Graham C. Wilson
Department of Primary Industries and Fisheries, Queensland

Revised version 2007:
Sergio Fernández
Ezequiel Rosales
Gilbert Weber
José-María Hernández
DSM Nutritional Products

Published by 5M Publishing
Benchmark House
8 Smithy Wood Drive
Sheffield, S35 1QN
Great Britain
www.thepoultrysite.com
books@5mpublishing.com

ISBN 0-9530150-6-8

Printed and published in the United Kingdom, February 2007.
Reprinted September 2010.

No part of this publication may be reproduced, copied or transmitted save with prior written permission or in accordance with the provisions of the Copyright Act 1956 (as amended), or under the terms of any licence permitting copying issued by the Copyright Licensing Agency, 90 Tottenham Court Rd, London W1T 4LP, Tel 020 7631 5555, Fax 020 7631 5500, email cla@cla.co.uk, web www.cla.co.uk

Any person who does any unauthorised act in relation to this publication may be liable to criminal prosecution and civil claims for damages.

A CIP catalogue record for this book is available from the British Library.

©The State of Queensland, Australia (through its Department of Primary Industries and Fisheries) and DSM Nutritional Products Ltd, 2007.

Enquiries should be addressed to:
copyright@dpi.qld.gov.au (phone: +61 7 3404 6999), or Director, Intellectual Property Commercialisation Unit, Department of Primary Industries and Fisheries, GPO Box 46 Brisbane, Queensland, Australia 4001

Contents

Preface 9
Formation of the egg 11
Optimum vitamin nutrition of laying hens 12
The nutritive value of the egg 14
Internal and external egg quality 15
The importance of calcium and vitamin D_3 for optimizing eggshell quality 17
Quality control 18
Changes in quality as the egg ages 19
Egg quality in the retail store and in the home 20
Consumer perceptions on egg quality 21
Egg defects 24
Nutrient Check List 26

Shell defects
Gross cracks 27
Hairline cracks 28
Star cracks 30
Thin-shelled eggs and shell-less eggs 32
Sandpaper or rough shells 34
Misshapen eggs 35
Flat-sided eggs 36
Body-checked eggs 37
Pimples 38
Pinholes 39
Mottled or glassy shells 40
Cage marks 41
Stained eggs 42
Fly marks 43
Fungus or mildew on shells 45
 46

Internal defects
Blood spots 47
Meat spots 48
Watery whites 49
Pale yolks 50
Mottled yolks and discoloured yolks 52
Discoloured whites 54
Rotten eggs 56
Roundworms in eggs 57
Off odours and flavours 59
 60

Glossary 63

Preface

Defects in hen eggs are a major concern for commercial producers and marketing agencies.

They result in a loss of industry efficiency and, should poor-quality eggs get through to the consumer, a loss of confidence in the product.

It is essential for both the industry and the consumer that the incidence of egg defects be minimised at all levels of production and marketing. Producers, in particular, must be able to quickly pinpoint and correct problems.

This handbook has drawn together the essential information on egg production and egg quality. It gives valuable information on the optimum nutrition of laying hens and explains the importance of vitamins and minerals for optimizing egg quality. Furthermore, egg defects are described in detail by:

- a description and a colour photograph
- the likely incidence in well-managed hens
- the possible causes
- solutions.

The book will be a valuable reference for all sectors of the egg industry: industry suppliers, food handling, preparing and processing industries, poultry fanciers, home egg producers, and students and teachers of poultry management.

We wish to acknowledge the valuable input and assistance from our colleagues in the Department of Primary Industries and Fisheries, Queensland (DPI & F) Poultry Unit, and the members of the National Coordinating Group for Poultry Research and Extension. Permission given by the Australian Egg Marketing Council to use material in sections of this publication is gratefully acknowledged.

We particularly wish to thank John Connor and Paul Kent, DPI & F, for their invaluable contributions to this handbook.

Formation of the egg

The egg is formed gradually over a period of about 25 hours. Many organs and systems help to convert raw materials from the food eaten by the hen into the various substances that become part of the egg.

The ovary

The hen, unlike most animals, has only one functional ovary — the left one — situated in the body cavity near the backbone. At the time of hatching, the female chick has up to 4000 tiny ova (reproductive cells), from some of which full-sized yolks may develop when the hen matures. Each yolk (ovum) is enclosed in a thin-walled sac, or follicle, attached to the ovary. This sac is richly supplied with blood.

The oviduct

The mature yolk is released when the sac ruptures, and is received by the funnel of the left oviduct (the right oviduct is not functional). The left oviduct is a coiled or folded tube about 80 cm in length. It is divided into five distinct sections, each with a specific function, as summarised in table 1.

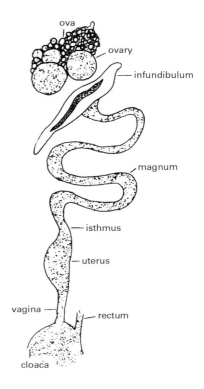

Figure 1: Reproductive organs of the hen

Table 1: Functions of various different sections of the hen's oviduct

Section of oviduct	Approximate time egg spends in this section	Functions of section of oviduct
1 Funnel (infundibulum)	15 minutes	Receives yolk from ovary. If live sperm present, fertilisation occurs here (commercially produced table eggs are not fertilised)
2 Magnum	3 hours	Inner and outer shell membranes are added, as are some water and mineral salts
3 Isthmus	1 hour	Albumen (white) is secreted and layered around the yolk
4 Shell gland (uterus)	21 hours	Initially some water is added, making the outer white thinner. Then the shell material (mainly calcium carbonate) is added. Pigments may also be added to make the shell brown
5 Vagina/cloaca	less than 1 minute	The egg passes through this section before laying. It has no other known function in the egg's formation

Optimum vitamin nutrition of laying hens

The overall goal of the layer industry is to achieve the best performance, feed utilization and health of birds. All nutrients including proteins, fats, carbohydrates, vitamins, minerals and water are essential for these vital functions, but vitamins have an additional dimension. They are required in adequate levels to enable the animal to efficiently utilize all other nutrients in the feed. Therefore, optimum nutrition occurs only when the bird is offered the correct mix of macro- and micronutrients in the feed and is able to efficiently utilize those nutrients for its growth, health, reproduction and survival.

Vitamins are active substances, essential for life of man and animals. They belong to the micronutrients and are required for normal metabolism in animals. Vitamins are essential for optimum health as well as normal physiological functions such as growth, development, maintenance and reproduction. As most vitamins cannot be synthesized by poultry in sufficient amounts to meet physiological demands, they must be obtained from the diet. Vitamins are present in many feedstuffs in minute amounts and can be absorbed from the diet during the digestive process. If absent from the diet or improperly absorbed or utilized, vitamins are a cause of specific deficiency diseases or syndromes.

Classically, vitamins have been divided into two groups based on their solubility in lipids or in water. The fat-soluble group includes the vitamins A, D, E and K, while vitamins of the B complex (B_1, B_2, B_6, B_{12}, niacin, pantothenic acid, folic acid and biotin) and vitamin C are classified as water-soluble. Fat-soluble vitamins are found in feedstuffs in association with lipids. The fat-soluble vitamins are absorbed along with dietary fats, apparently by mechanisms similar to those involved in fat absorption. Water-soluble vitamins are not associated with fats, and alterations in fat absorption do not affect their absorption, which usually occurs via simple diffusion. Fat-soluble vitamins may be stored in the animal body. In contrast, water-soluble vitamins are not stored, and excesses are rapidly excreted.

It is now well recognized by the feed industry that the minimum dietary vitamin levels required to prevent clinical deficiencies may not support optimum health, performance and welfare of poultry. The reasons for this are manifold: The productivity of poultry farming continues to grow through genetic improvement of the breeds and through modifications in nutrition, management and husbandry, which considerably increase the demand for vitamins. Furthermore, intensive poultry production may generate a certain level of metabolic, social, environmental and disease stresses, causing sub-optimal performance and higher susceptibility to vitamin deficiencies. The contamination of the feed with mycotoxins and vitamin antagonists can limit or even block the action of certain vitamins. Any of these factors, ranging from the animals' genetic background and health status to management programmes and the composition of the diet, can separately or collectively affect the need for each vitamin. As intake and utilization of vitamins from natural sources is unpredictable owing to differing contents of vitamins in the feedstuffs (dependent on growing climate and harvesting time of crops, processing and storage conditions of feed ingredients) and variable vitamin bioavailability, it is safer to cover the total vitamin requirement of poultry through dietary supplementation.

More than ever before, the layer industry is currently facing the challenge to improve productivity in order to remain competitive in today's cost-driven environment. Fortunately, high-performing layer breeds with improved performance pattern, optimized feed conversion capabilities and favourable health characteristics are available. But in order to allow the birds to perform up to their genetic potential, their nutrition and especially their vitamin supply needs to be optimized. In particular, B vitamins are required for efficient nutrient utilization, and together with vitamin A are important to support the hens' metabolic activity for maintenance and high laying performance. Furthermore, both vitamins C and E improve the birds' resistance to stress, and help sustain health and longevity. Specific benefits related to superior egg quality can be achieved if supra-nutritional levels of vitamin E are added to the feed of laying hens. And finally, considerable vitamin D activity is required in order to support an adequate skeletal development and to avoid leg problems of various origins.

The optimum vitamin supplementation levels are given in the table below.

Vitamins (added to air-dried feed)	Replacement pullets	Laying hens
Vitamin A (IU/kg)	7 000–10 000	8 000–12 000
Vitamin D_3 (IU/kg)	1 500–2 500	2 500–3 500[1]
Vitamin E (mg/kg)	20–30	15–30[2]
Vitamin K_3 (mg/kg)	1–3	2–3
Vitamin B_1 (mg/kg)	1.0–2.5	1.5–3.0
Vitamin B_2 (mg/kg)	4–7	4–7
Vitamin B_6 (mg/kg)	2.5–5.0	3.0–5.0
Vitamin B_{12} (mg/kg)	0.015–0.025	0.015–0.025
Niacin (mg/kg)	25–40	20–50
Pantothenic acid (mg/kg)	9–11	8–10
Folic acid (mg/kg)	0.8–1.2	0.5–1.0
Biotin (mg/kg)	0.10–0.15	0.10–0.15
Vitamin C (mg/kg)	100–150	100–200
Hy•D® (25-OH D_3) (mg/kg)	0.069[3]	0.069[3]
Choline (mg/kg)	200–400	300–500

[1] Do not exceed 3000 IU D3/kg feed when using Hy•D®
[2] Under heat stress conditions: 200 mg/kg
[3] Local legal limits of total dietary vitamin D activity need to be observed
Source: DSM Vitamin Supplementation Guidelines, 2006; Optima Nutrición Vitamínica de los animales para la producción de alimentos de calidad, 2002

The nutritive value of the egg

The egg is one of the most complete and versatile foods available. It consists of approximately 10% shell, 58% white and 32% yolk. Neither the colour of the shell nor that of the yolk affects the egg's nutritive value. The average egg provides approximately 313 kilojoules of energy, of which 80% comes from the yolk.

The nutritive content of an average large egg (containing 50 g of edible egg) includes:

6.3 g protein
0.6 g carbohydrates
5.0 g fat (this includes 0.21 g cholesterol).

Egg protein is of high quality and is easily digestible. Almost all of the fat in the egg is found in the yolk and is easily digested.

Vitamins

Eggs contain every vitamin except vitamin C. They are particularly high in vitamins A, D, and B_{12}, and also contain B_1 and riboflavin. Provided that laying hens are supplemented according to the Optimum Vitamin Nutrition concept (see chapter 'Optimum vitamin nutrition of laying hens'), eggs are an important vehicle to complement the essential vitamin supply to the human population.

Minerals

Eggs are a good source of iron and phosphorus and also supply calcium, copper, iodine, magnesium, manganese, potassium, sodium, zinc, chloride and sulphur. All these minerals are present as organic chelates, highly bioavailable, in the edible part of the egg.

Internal and external egg quality

Quality has been defined by Kramer (1951) as the properties of any given food that have an influence on the acceptance or rejection of this food by the consumer. Egg quality is a general term which refers to several standards which define both internal and external quality. External quality is focused on shell cleanliness, texture and shape, whereas internal quality refers to egg white (albumen) cleanliness and viscosity, size of the air cell, yolk shape and yolk strength.

Internal egg quality

Internal egg quality involves functional, aesthetic and microbiological properties of the egg yolk and albumen. The proportions of components for fresh egg are 32% yolk, 58% albumen and 10% shell (Leeson, 2006).

The egg white is formed by four structures. Firstly, the chalaziferous layer or chalazae, immediately surrounding the yolk, accounting for 3% of the white. Next is the inner thin layer, which surrounds the chalazae and accounts for 17% of the white. Third is the firm or thick layer, which provides an envelope or jacket that holds the inner thin white and the yolk. It adheres to the shell membrane at each end of the egg and accounts for 57% of the albumen. Finally, the outer thin layer lies just inside the shell membranes, except where the thick white is attached to the shell, and accounts for 23% of the egg white (USDA, 2000).

Egg yolk from a newly laid egg is round and firm. As the egg gets older, the yolk absorbs water from the egg white, increasing its size. This produces an enlargement and weakness of the vitelline membrane; the yolk looks flat and shows spots.

As soon as the egg is laid, its internal quality starts to decrease: the longer the storage time, the more the internal quality deteriorates. However, the chemical composition of the egg (yolk and white) does not change much.

In a newly laid egg the albumen pH lies between 7.6 and 8.5. During storage, the albumen pH increases at a temperature-dependent rate to a maximum value of about 9.7 (Heath, 1977). After 3 days of storage at 3 °C, Sharp and Powell (1931) found an albumen pH of 9.18. After 21 days of storage, the albumen had a pH close to 9.4, regardless of storage temperature between 3 and 35 °C (Li-Chan et al, 1995).

Heath (1977) observed that when carbon dioxide (CO_2) loss was prevented by the oiling of the shell, the albumen pH of 8.3 did not change over a 7-day period of storage at 22 °C. In oiled eggs stored at 7 °C, albumen pH dropped from 8.3 to 8.1 in seven days (Li-Chan et al, 1995).

Increases in albumen pH are due to CO_2 loss through the shell pores, and depend on dissolved CO_2, bicarbonate ions, carbonate ions and protein equilibrium. Bicarbonate and carbonate ion concentration is affected by the partial CO_2 pressure in the external environment.

In newly laid eggs, the yolk pH is in general close to 6.0; however, during storage it gradually increases to reach 6.4 to 6.9. Egg quality preservation through handling and distribution is dependent on constant care from all personnel involved in these activities. The quality of the egg once it is laid cannot be improved, so efforts to maintain its quality must start right at this moment.

The decrease in internal egg quality once the egg is laid is due to the loss of water and CO_2. In consequence, the egg pH is altered, resulting in watery albumen due to the loss of the thick albumen protein structure. The cloudy appearance of the albumen is also due to the CO_2; when the egg ages, the CO_2 loss causes the albumen to become transparent, compared with fresh eggs.

To minimize egg quality problems two things are important: frequent egg collection, mainly in the hot months, and rapid storage in the cool room. The best results are obtained at a temperature of 10 °C. There are six main factors affecting internal egg quality: disease, egg age, temperature, humidity, handling, and storage.

Disease: Newcastle disease and infectious bronchitis produce watery albumen, and this condition may persist for long periods after the disease outbreak has been controlled (Butcher, 2003).

Egg age: eggs several days old show weak and watery albumen, and the CO_2 loss makes the content alkaline, affecting the egg flavour.

Temperature: high temperatures cause a rapid decrease in internal quality. Storage above 15.5 °C increases humidity losses.

Humidity: high relative humidity (RH) helps to decrease egg water losses. Storage at an RH above 70% helps to reduce egg weight losses and keeps the albumen fresh for longer periods of time.

Egg handling: rough handling of the eggs not only increases the risk of breaking the eggs, but also may cause internal egg quality problems.

Storage: eggs are very prone to take on the odours of other products stored with them; separate storage is therefore advised.

The variables mentioned above are particularly important to ensure that a 1-week-old egg, properly handled, can be as fresh as a day-old egg kept at room temperature.

If the egg is properly handled during shipment and distribution, it will reach the consumer's table with adequate freshness.

External egg quality

Poor eggshell quality has been of major economic concern to commercial egg producers, with estimated annual losses in the USA of around 478 million US dollars (Roland 1988). In Australia in 1998, the impact was of the order of 10 million Australian dollars per year. Information obtained from egg grading facilities indicates that 10% of eggs are downgraded due to egg shell quality problems. Based on values for the UK, Germany and the USA, it has been estimated that the incidence of broken eggs ranges between 6 and 8% (Washburn, 1982). In Mexico in 2005 it was estimated that the egg industry lost between 30 and 35 million US dollars, based on average figures of 2.5% broken eggs and 4% weak shells. These losses occur only between laying and packing, not taking into account losses in transit to the end consumer (DSM Mexico, 2005, unpublished data).

To maintain consistently good shell quality throughout the life of the hen, it is necessary to implement a total quality management programme throughout the egg production cycle.

It has been always recognized that the hen has the most extraordinary method of obtaining and depositing calcium (Ca) in the entire animal kingdom. An egg has an average of 2.3 g of calcium in the shell, and almost 25 mg in the yolk (Etches, 1987). A modern hen laying 330 eggs per cycle will deposit 767 g of calcium; assuming a 50% calcium retention rate from the diet, the hen will consume 1.53 kg of calcium per cycle.

Exterior egg quality is judged on the basis of texture, colour, shape, soundness and cleanliness according to USDA (2000) standards. The shell of each egg should be smooth, clean and free of cracks. The eggs should be uniform in colour, size and shape.

There are five major types of shell problems in the egg industry: 1. cracks due to excess pressure; 2. cracks due to thin shells; 3. body-checks; 4. pimpled or toe holes, and 5. shell-less eggs.

When a producer complains about an increase in downgrade eggs, the first thing required is to determine which types of problems have increased. In a processing plant with 97% A-quality eggs, a typical distribution of the different types of shell problems (downgrade) might be 2.13% stains, 0.85% blood spots, 0.85% meat spots, 61% pressure cracks, 9.8% thin shell cracks, 6.8% body-checks, 13.6% pimpled and 5.1% toe holes. If the percentage of any type of shell problem is abnormally high, then that is the problem needing attention.

The importance of calcium and vitamin D_3 for optimizing eggshell quality

The eggshell is a highly specialized mineralized structure, which provides protection against physical damage and penetration by micro-organisms. The egg shell consists of the inner and outer shell membranes, the true shell and the cuticle. The crystalline layer of the shell, which is responsible for its mechanical strength, consists of more than 90% calcium in the form of calcium carbonate. Calcium is absorbed from the feed in the intestine. Provided that sufficient calcium (3.8–4.2%) is present in the feed, the process of calcium uptake, deposition and excretion is regulated by vitamin D_3 and its metabolites.

Vitamin D_3 is absorbed from the intestine in association with fats and requires the presence of bile salts for absorption. It is transported via the portal circulation to the liver, where it is accumulated. The first transformation occurs in the liver, where vitamin D_3 is hydroxylated to become 25-hydroxyvitamin D_3 (25-OH D_3). This vitamin D_3 metabolite is then transported to the kidney where it is converted to the most active hormonal compound 1,25-dihydroxyvitamin D_3 (1,25-$(OH)_2$ D_3). The production of 1,25-$(OH)_2$ D_3 is tightly regulated by parathyroid hormone (PTH) in response to serum calcium. If plasma calcium is low, PTH secretion is induced, which stimulates the hydroxylation of 25-OH D_3 to 1,25-$(OH)_2$ D_3. This compound will increase calcium absorption in the intestine, mobilize calcium from the bones and reduce calcium excretion via the kidney. If plasma calcium is high, first PTH secretion and then 1,25-$(OH)_2$ D_3 production are suppressed, which result in a reduction of calcium absorption in the gut as well as calcium resorption from the bones and an increase in calcium excretion. Therefore it is of utmost importance for an optimum egg shell quality to optimize calcium supply and secure sufficient vitamin D_3 activity available to the laying hen.

Despite adequate fortification of layer feeds with vitamin D_3, clinical signs of vitamin D_3 deficiency such as rickets or cage-layer-fatigue can frequently be observed in laying hens kept under commercial conditions. Such disorders indicate insufficient utilization of the dietary vitamin D_3, which can be counteracted by a special feed product such as ROVIMIX® Hy•D®.

Chemically, Hy•D® is 25-OH D_3, representing the first metabolite in the cascade of vitamin D mobilization. In numerous studies, Hy•D® has been demonstrated to support the homeostatic function of vitamin D_3, which is important to provide sufficient minerals for incorporation into the bone matrix as well as for optimizing the stability of the eggshell. Hy•D® also helps to maximize bone mass before the onset of lay and thus prevents layers from a fatal demineralization of the bones, resulting in osteoporosis. Therefore, Hy•D® is an effective and more flexible source of vitamin D_3 activity for optimizing vitamin nutrition and ultimately maximizing profitability of modern layer production.

Quality control

Eggs for the fresh, in-shell market must meet strict standards to ensure that only those of high quality reach the consumer.

The first step in the quality control process is the segregation of eggs with obvious defects. The major technique used in subsequent quality tests on the rest of the eggs is candling. In this process, all eggs are passed over a bright light which shows up internal defects and previously undetected cracked or weak shells. The internal defects often detectable by candling include blood and meat spots, enlarged air cells and very thin whites.

Samples of eggs are also taken to assess egg freshness and yolk colour. These eggs are broken out onto a level surface and the height of the thick albumen is measured with a micrometer. From this measurement and the weight, the Haugh unit value of an egg can be calculated. The fresher the egg, the higher its Haugh unit value.

Yolk colour is checked against a required colour standard, the DSM Yolk Colour Fan (formerly known as the Roche Yolk Colour Fan) being the one used by the egg industry worldwide.

Although definitions may differ between areas, in general, on candling first-quality hen eggs must be fresh and free from blood spots and other inclusions, and must not have been subjected to incubation. The shell must be clean, sound and not misshapen.

The main difference between second-quality and first-quality eggs is that the shells of second-quality eggs may be cracked and misshapen, but the shell membrane must be intact. They may also be less fresh, which is indicated by an enlarged air cell and watery white, but must be free from rot. Second-quality eggs, appropriately identified, may be sold (but not re-sold) in the shell for human consumption, or may be used for the production of egg pulp (bulk egg content mainly used in the food industry).

Eggs which do not conform to the requirements for first-quality or second-quality eggs are not fit for human consumption. Examples of such eggs are those with large blood spots or other inclusions, rotten eggs and those with ruptured shell membranes.

Changes in quality as the egg ages

Changes in quality as the egg ages are summarised in figure 2. To slow down these changes, new-laid eggs can be put in cool storage, and/or the shells covered with a thin layer of an approved oil, particularly over the air cells.

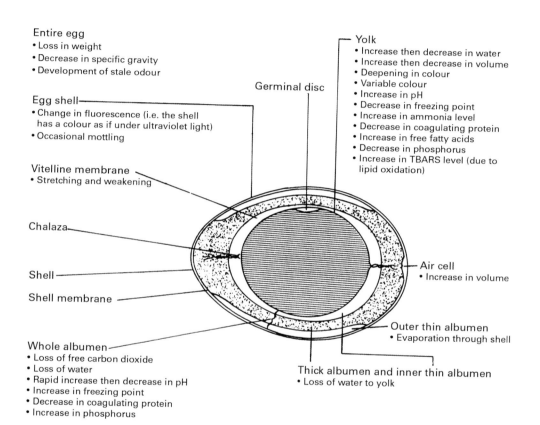

Entire egg
- Loss in weight
- Decrease in specific gravity
- Development of stale odour

Egg shell
- Change in fluorescence (i.e. the shell has a colour as if under ultraviolet light)
- Occasional mottling

Vitelline membrane
- Stretching and weakening

Chalaza

Shell

Shell membrane

Whole albumen
- Loss of free carbon dioxide
- Loss of water
- Rapid increase then decrease in pH
- Increase in freezing point
- Decrease in coagulating protein
- Increase in phosphorus

Germinal disc

Yolk
- Increase then decrease in water
- Increase then decrease in volume
- Deepening in colour
- Variable colour
- Increase in pH
- Decrease in freezing point
- Increase in ammonia level
- Decrease in coagulating protein
- Increase in free fatty acids
- Decrease in phosphorus
- Increase in TBARS level (due to lipid oxidation)

Air cell
- Increase in volume

Outer thin albumen
- Evaporation through shell

Thick albumen and inner thin albumen
- Loss of water to yolk

Figure 2: Structure of the egg. A summary is given of the changes occurring as the egg ages.

Egg quality in the retail store and in the home

This handbook deals mainly with the producer's influence on egg quality. Retailers and consumers can also minimise the decline in egg quality by observing some basic guidelines.

The way eggs are stored is as important as how long they are stored. Eggs can lose as much quality in one day at room temperature as in 4 to 5 days in the refrigerator. Retailers should have adequate refrigerated holding space, avoid storage of eggs close to strong-smelling foods, display eggs away from sunlight, and strictly rotate stocks (i.e. keep older stock in front of fresh stock so that all eggs are sold as fresh as possible).

Consumers should avoid handling eggs roughly and leaving them in hot vehicles. At home, eggs are best stored in the refrigerator in their carton. The carton provides protection from damage, slows down moisture loss and helps prevent the absorption of odours from strong-smelling foods.

Consumer perceptions of egg quality

The European Consumer Association (BEUC) has indicated some of the attributes (food quality factors) that are valued by consumers but, until recently, there had been little detailed information published specifically regarding consumer perception of egg quality. However, data from consumer surveys performed over recent years has added greatly to our knowledge in this area.

In 2001, a study involving 3085 people was performed in Spain with the specific objective of validating and ranking in eggs those attributes proposed by the BEUC. Not surprisingly, the survey showed that for consumers, 'safety' and 'freshness' were the most important quality factors, with 'nutritional value' and 'sensory characteristics' also being key parameters. With regards to 'sensory characteristics', the results of surveys performed over the last 10 years in a number of European countries (France, Germany, Italy, UK, Spain, Poland and Greece) indicate that consumers value a number of tangible characteristics of the egg, most especially shell strength, albumen consistency and yolk colour.

Yolk colour

Yolk colour, both in terms of the colour per se and the variability in colour among eggs, is a very important parameter by which consumers judge the quality of eggs.

In the surveys, when offered samples of eggs with different yolk colours (measuring 8, 10, 12 and 14 on the DSM Yolk Colour Fan (YCF), the majority of the people questioned in all countries expressed a preference for the egg yolks with the darkest colour hue (colour score 14). Yolk colour in laying hens is primarily determined by the content and profile of pigmenting carotenoids present in their feed and can easily be adapted via feed ingredients to match consumer preferences.

In this regard, it is surprising to note that in data published in 2004 by an independent quality control laboratory auditing eggs for supermarkets in Spain (analysis of 12 000 eggs during 2002/3), yolk colour showed the highest non-compliance to egg quality specifications (8.8% eggs below specification). There are several explanations as well as some potential solutions which may help minimize this effect. In many cases, egg producers understand the target yolk colour as the average yolk colour they need to obtain.

Taking the example of an egg producer with a target DSM-YCF score of 12 (tolerance -1) in their retail specification, by producing eggs to the target (DSM-YCF score 12), a percentage of the eggs produced will fail against the retail specification due to normal biological variation.

The most straightforward way to counteract this and avoid such a high level of non-compliances is to increase the average yolk colour of the total egg production. In addition to inherent biological variation, it is also important to minimise the effect of other factors which can also reduce the consistency of the yolk colour produced — the wider the distribution of yolk colours about the mean value, the higher will be the potential for non-compliance with the retailer specification.

For example, in order to achieve both the target level and homogenous distribution of the carotenoids in the feed, it is important to use reliable, uniform products which have both good stability (commercial product forms/premix/feed) and mixability characteristics.

Likewise, it is also important to provide an adequate level of yellow carotenoids in the feed to provide a good yellow base for the development of the required golden yellow colour in the egg yolk. By achieving the correct levels of the most appropriate carotenoids in the feed, it is possible to routinely achieve the yolk colour expected by consumers while at the same time minimizing the degree of variability and thereby reducing the potential for non-compliance.

www.yellow-egg.com website

How can you tell if an egg is fresh? What does the stamp on the egg mean? Do brown eggs taste better? What does the colour of the egg yolk say about the hen? What tasty dishes can I cook up using eggs? Are you able to explain to your kids how a hen makes an egg?

Now there is a website, www.yellow-egg.com, accessible to everyone in the food chain to try to help you find most answers. The yellow-egg web site is currently available in English, Spanish, Italian, German, French and soon in Russian, Portugese, Polish and Japanese.

Why are egg yolks yellow?

Food doesn't just have to taste good, it has to look good too: blue milk, green meat, and violet butter don't stand a chance under our critical gaze. Everything that we eat has to look good, smell good, taste good, and have the right consistency. Especially when it comes to egg yolks, we love a golden yellow. But where does this golden yellow come from? What are carotenoids, and what role do they play in nature?

What are carotenoids?

In 1931, the chemist Heinrich Wackenroder was the first to discover a carbon compound in carrots, naming it "carotene". We now know of some 650 carotenoids, without which natural life would be impossible. These substances are responsible, for example, for the yellow to reddish-orange colours of fruit and vegetables, for the vibrant hues of flowers and for many a colourful coat in the animal kingdom. Plants, fungi and bacteria around the world produce some three tons of carotenoids every single second.

In nature, carotenoids are much more than just colourants. They also perform vital protective and regulatory functions. Neither humans nor animals are capable of producing carotenoids themselves; we need to take them in with our food. Around fifty of these valuable colorants are important for humans as pro-vitamin A, meaning that they can be converted into vitamin A. Some of carotenoid's biological functions are:

- Protect the cells of the body from harmful environmental influences. (e.g "free radicals")
- Improve the performance of the immune system.
- Support detoxifying functions.
- Are involved in the process by which we see.
- Protect the skin from damage by ultraviolet light.
- Increase the fertility of animals.

The SQTS Concept

SQTS means the way chosen by leading companies to manufacture their feed ingredient portfolio, carotenoids being a good example of it: the best way to guarantee safety, quality, traceability and sustainability to the Food Chain.

"S" for Safety
Hen feed ingredients are manufactured according to the highest possible standards and production is subject to strict monitoring. Of course, this also applies to carotenoids. It sees to the hen's health and gives the egg its "heart of gold" — a golden yellow yolk.

"Q" for Quality
Globally, around one thousand billion eggs are consumed every year. Numerous consumer surveys show that the colour of the yolk is a distinct sign of quality. Consumers want eggs with consistently coloured egg yolks. We first notice yolk quality at the breakfast table, but it is around long before that: through hen feed enriched with carotenoids the farmer is able to influence the homogeneity and intensity of the yolk colour.

"T" for Traceability
In the EU every egg is individually stamped. The codes are standardized throughout the EU and allow the egg to be traced through an unbroken chain back to the farm where it was laid. This protects

consumers and egg farmers in equal measure. Transparency in the entire food chain guarantees optimum safety for egg consumers. This high standard of safety also applies to responsible companies producing feed ingredients for the egg chain. Every product and feed ingredient must be traced back to its origin.

"S" for Sustainability
Just as an example, did you know that in order to ensure the global need for vitamin C, 30% of the earth would have to be covered by orange groves? Sustainable agriculture secures our standard of living – without devastating our natural reserves. The same would happen with any other feed or food ingredient identical to those found in nature and manufactured using the most modern production methods. This conserves the environment and ensures us a comfortable and healthy life. Sustainable agriculture, for the environment's sake.

Egg defects

In this book, individual egg defects are discussed in two groups: shell defects and internal defects. Shell defects are discussed first, as they are the most common cause of downgrading. Individual defects are discussed in general order of decreasing frequency of occurrence within each group. Each defect is described and illustrated, and its expected incidence at the point of grading is given for comparison where appropriate. (The given incidence is for eggs produced under conditions of good management.) Possible causes of an excessive incidence of each defect and the corresponding control measures are outlined.

To put an egg-quality problem occurring on a farm into the correct perspective, it is important to go through the following steps.

1. Quantify the extent of the farm problem
Accurately determine the incidence or extent of the problem. Count all eggs with the problem, both at collection and at grading.

2. Determine flock age
Flock age has a major impact on the incidence of both external and internal egg defects and so must be taken into account when assessing the severity of a problem. If the birds on the farm vary in age, calculate the weighted average age of the flock, as shown in table 2.

The graph in figure 3 shows the relationship between flock age and both the incidence of second-quality eggs and internal egg quality (measured in Haugh units).

An increased incidence of defects may be primarily the result of an increase in the average age of the laying flock.

3. Assess grading efficiency
If too many poor-quality eggs are being packed for sale, the problem may lie in the grading-candling operation or subsequent rough handling. Factors that affect the efficiency of the grading-candling operation include:

Candling speed. As the time that the candling operator has to view each egg decreases (for example, from 0.6 seconds per egg to 0.2 seconds per egg), candling efficiency declines.

Proportion of second-quality eggs. As the percentage of second-quality eggs entering the candling booth increases (for example, beyond 8 to 10%), candling efficiency declines. Removing obviously poor-quality eggs before grading will help reduce the workload of candling operators.

Candling light. Although the intensity of light in itself is not critical, leakage of light around the eggs may reduce efficiency.

Candling mirror. Backing mirrors that are not adjusted to the eye-height of the candling operator will decrease candling efficiency.

Operator. The experience and capability of the operator is a major factor affecting efficiency. Operators should be able to stop the line when overloaded, and should not work at candling too long without a break.

Table 2: Sample calculation of weighted average age of flock

Flock number	Number of birds in lay in each flock		Age of each flock (weeks)		Total number of birds in lay in all flocks		
1	1000	×	25	÷	5050	=	4.9 +
2	1200	×	32	÷	5050	=	7.6 +
3	1550	×	47	÷	5050	=	14.4 +
4	1300	×	64	÷	5050	=	16.4
Total	5050				Average flock age (weeks)		43.5

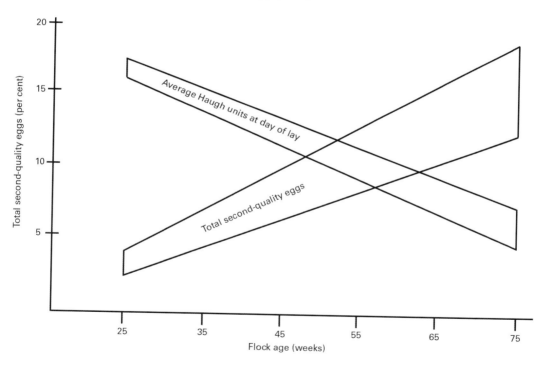

Figure 3: The relationship between flock age and the incidence of second-quality eggs, and the Haugh unit measure of internal egg quality.

By systematically following the above three steps, an apparent egg-quality problem can be quantified and put into perspective. If the incidence of a defect is too high, possible causes and solutions can be pinpointed by consulting the section of this book on the defect concerned.

Nutrient check list

Nutritional factors are often suspected as a prime cause of shell-quality problems. If such problems occur, check the following list of nutrients which play an important role in shell quality:

Calcium
Calcium intake should be between 3.8 and 4.2 g per bird and per day, and be maintained by adjusting the diet formulation or by the use of calcium 'icing' (sprinkling large particles of a calcium source on top of feed in troughs), starting 2 weeks before lay.

Phosphorus
Total phosphorus intake should be below 0.8 g per bird and per day. The ratio of calcium to phosphorus should be as near 6:1 as practical.

Vitamin D_3
Most custom-made premixes are adequate (2500–3500 IU/kg feed) in this vitamin, but high storage temperatures, long periods in storage and poor mixing may result in deficiencies. Products such as ROVIMIX® Hy•D® might be considered to complement the vitamin D_3 supply of laying hens.

Magnesium
Problems can arise if intakes are greater than 0.7 g per bird and per day. Calcium sources that contain magnesium, particularly dolomitic limestones, can result in magnesium levels that are too high.

Chloride
An excess of chloride in hot weather may increase shell-quality problems. Dietary chloride should not exceed 0.25%.

Saline water
Relatively low levels of excess salts in drinking water can increase the incidence of cracked and broken eggs within one week of exposure. For example, the ability of the bird to produce normal egg shells is permanently affected by levels of sodium chloride from as low as 200 mg/L.

Note that the actual intake of the nutrient is important, not just the proportion of a nutrient in the diet. For example, if calcium is included in a diet at 4.0%, but the hen is consuming only 90 g of feed per day, its actual intake is 4.0/100 x 90 = 3.6 g of calcium per day. Intake can be affected by the energy level of the diet, the breed of hen, the stage of production, and the environmental temperature. (High shed temperatures can also decrease the efficiency with which the hen uses the nutrients she does consume.)

Shell defects

Gross cracks

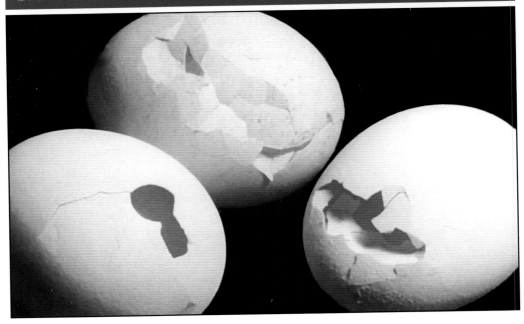

Description
The term 'gross cracks' refers to large cracks and holes, which usually result in a broken shell membrane.

Incidence
The incidence of gross cracks increases with the age of the hen. It ranges from 1 to 5% of total production.

Cause	Control
Reduced shell strength due to:	
Ageing	Keep flock age as low as economically possible.
Poor nutrition	Make sure that the birds' nutrient intake is correct (particularly regarding calcium and vitamin D_3). Mixed feed should be handled carefully so that the different components do not separate out. This particularly needs to be checked when augers and automatic feeding systems are used.
Saline water	Desalinate, dilute or do not use drinking water containing problem levels of salts.
Diseases such as infectious bronchitis	Follow an effective vaccination programme.
High shed temperatures	Control temperatures by using foggers, fans, roof sprinklers, white roofs, insulation and good ventilation.

(continued)

Cause	Control
Mechanical damage caused by birds' beaks and toenails	Control egg eating by the birds. Make sure that birds do not have access to eggs in roll-out trays. Use cages designed to prevent access. Reduce damage caused by the birds' toenails. The slope and construction of the cage floor should allow the eggs to move freely to the roll-out tray.
Infrequent egg collection	Collect eggs at least twice a day.
Rough handling	Do not collect eggs in wire baskets. When stacking fillers of eggs, place one empty filler at the bottom of the stack, and a full one directly on top of it. This double bottom layer supports the weight of the stack better. Avoid stacking fillers of eggs more than six high. Pick up and carry stacks with care. Place large eggs on top fillers in a stack. Reduce the severity of impacts during mechanical handling by: • cushioning metal egg guides; • keeping egg roll-out angles between 7 and 8°; • minimising the number of rows of eggs being fed onto cross-conveyor belts at any one time. Educate staff to handle eggs with care during collection and packing.

Hairline cracks

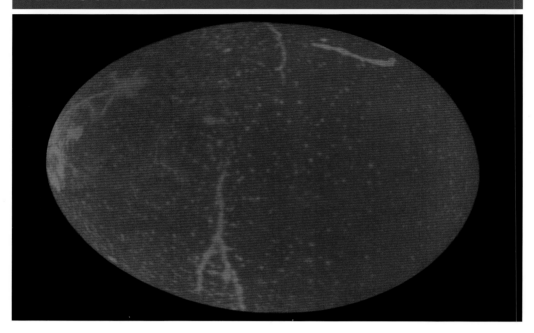

Description
Hairline cracks, i.e. very fine cracks, usually run lengthwise along the shell. As they are difficult to detect, candling efficiency needs to be maximised (see page 24). Their presence in fresh eggs can be revealed by careful squeezing or tapping. The crack becomes more obvious as the egg ages.

The egg in the photograph has been placed over a bright light. This is the way it would look to a candling operator.

Incidence
The incidence of this problem varies with flock age, but is usually 1 to 3% of total production.

Cause	Control
Reduced shell strength due to:	
Ageing	Keep flock age as low as economically possible.
Poor nutrition	Make sure that the birds' nutrient intake is correct (particularly regarding calcium and vitamin D_3) Mixed feed should be handled carefully so that the different components do not separate out. This particularly needs to be checked when augers and automatic feeding systems are used.
Saline water	Desalinate, dilute or do not use drinking water containing problem levels of salts.
Diseases such as infectious bronchitis	Follow an effective vaccination programme.
High shed temperatures	Control temperatures by using foggers, fans, roof sprinklers, white roofs, insulation and good ventilation.

(continued)

Cause	Control
Egg collisions or pressure on the egg due to poor design and maintenance of the cage floor	Make sure that: • the cage floor is not too rigid. • the slope of the floor is just enough to allow eggs to roll out. • roll-out trays cushion the eggs as they come to rest against the edge of the tray.
Infrequent egg collection	Collect eggs at least twice a day.
Rough handling	Do not collect eggs in wire baskets. When stacking fillers of eggs, place one empty filler at the bottom of the stack, and a full one directly on top of it. This double bottom layer supports the weight of the stack better. Avoid stacking fillers of eggs more than six high. Pick up and carry stacks with care. Place large eggs on top fillers in a stack. Reduce the severity of impacts during mechanical handling by: • cushioning metal egg guides. • keeping egg roll-out angles between 7 and 8°. • minimising the number of rows of eggs being fed onto cross-conveyor belts at any one time. Educate staff to handle eggs with care during collection and packing.

Star cracks

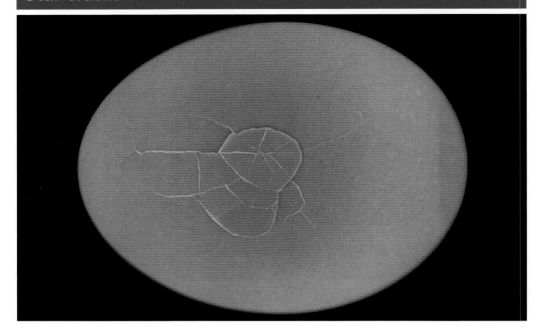

Description
Star cracks are fine cracks radiating outwards from a central point of impact, which is often slightly indented.

The egg in the photograph has been placed over a bright light. This is the way it would look to a candling operator.

Incidence
The incidence varies with flock age but is usually 1 to 2% of total production.

Cause	Control
Reduced shell strength due to:	
Ageing	Keep flock age as low as economically possible.
Poor nutrition	Make sure that the birds' nutrient intake is correct (particularly regarding calcium and vitamin D_3) Mixed feed should be handled carefully so that the different components do not separate out. This particularly needs to be checked when augers and automatic feeding systems are used.
Saline water	Desalinate, dilute or do not use drinking water containing problem levels of salts.
Diseases such as infectious bronchitis	Follow an effective vaccination programme.
High shed temperatures	Control temperatures by using foggers, fans, roof sprinklers, white roofs, insulation and good ventilation.

(continued)

Cause	Control
Egg collisions or pressure on the egg due to poor design and maintenance of the cage floor	Make sure that: • the cage floor is not too rigid. • the slope of the floor is just enough to allow eggs to roll out. • roll-out trays cushion the eggs as they come to rest against the edge of the tray.
Infrequent egg collection	Collect eggs at least twice a day.
Rough handling	Do not collect eggs in wire baskets. Reduce the severity of impacts during mechanical handling by: • cushioning metal egg guides. • keeping egg roll-out angles between 7 and 8°. • minimising the number of rows of eggs being fed onto cross-conveyor belts at any one time. Educate staff to handle eggs with care during collection and packing.

Thin-shelled eggs and shell-less eggs

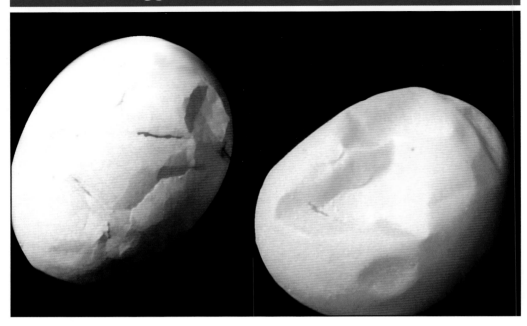

Description
Eggs with very thin shells, or no shell at all around the shell membrane, look unattractive and are highly susceptible to damage.

Incidence
The incidence of these eggs varies from about 0.5 to 6%. They are commonly produced by pullets coming into lay, particularly by birds that have matured early. Some birds continue to lay this type of egg.

Causes	Control
Immature shell gland	Delay onset of sexual maturity 1 to 2 weeks by controlled feeding during rearing.
Defective shell gland	Cull birds which persistently produce such eggs.
Disturbances causing eggs to be laid before calcification of the shell is complete	Minimise activities which create disturbances in and around the layer shed. Increase shed security to stop other birds and animals entering the shed.
Poor nutrition	Make sure that birds' nutrient intake is correct (particularly regarding calcium and vitamin D_3). Mixed feed should be handled carefully so that the different components do not separate out. This particularly needs to be checked when augers and automatic feeding systems are used.
Saline water	Desalinate, dilute or do not use drinking water containing problem levels of salts.
Diseases, e.g. infectious bronchitis and eggdrop syndrome	Follow effective vaccination programmes where available.

Sandpaper or rough shells

Description
The terms 'sandpaper shells' and 'rough shells' refer to eggs with rough-textured areas, often unevenly distributed over the shell.

Incidence
The incidence is normally less than 1% of total production, but may be higher for some strains of bird. It is also higher in early lay, often as a result of double ovulation, which produces one shell-less egg and another one with extra shell deposits.

Cause	Control
Diseases, e.g. infectious bronchitis, infectious laryngotracheitis or avian encephalomyelitis	Follow effective vaccination programmes.
Disturbances at the time a hen is due to lay can cause the egg to be held over for another day	Minimise activities which create disturbances in and around the shed. Increase shed security to stop other birds and animals entering the shed.
Incorrect or changes in lighting programme	There should not be sudden increases in day length as pullets come into lay, or lighting changes during lay.
Water shortages	Make sure that the water supply is adequate, that there are no blockages in water lines and that drinkers are functioning properly.

Misshapen eggs

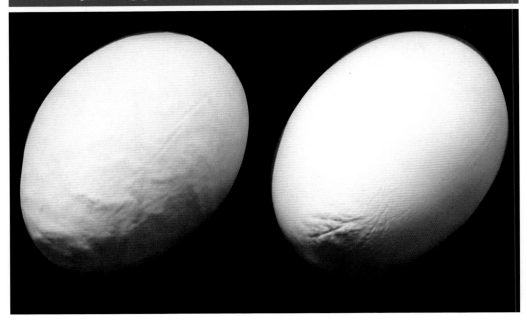

Description
Misshapen eggs are those whose shells differ obviously from the smooth, 'normal' shape. They include eggs with shells marred by flat sides or body checks (ribs or grooves), and eggs that are too large or too round. Flat sided eggs are discussed on page 37, and body-checked eggs on page 38.

Incidence
The incidence of misshapen eggs varies considerably depending on how severely the eggs are judged. Normally, up to 2% of production is downgraded due to these faults. The incidence of misshapen eggs can vary with the strain of bird, but they are most often produced by pullets coming into lay, or hens late in lay. often as a result of double ovulation.

Cause	Control
Immature shell gland	Delay onset of sexual maturity 1 to 2 weeks by controlled feeding during rearing.
Defective shell gland	Cull birds which persistently produce such eggs.
Diseases, e.g. infections bronchitis	Follow an effective vaccination programme.
Stress, e.g. frights and disturbances	To avoid frightening birds, minimise human activity in and around the shed. Increase shed security to stop other birds and animals entering the shed.
Crowding	Avoid overstocking.

Flat-sided eggs

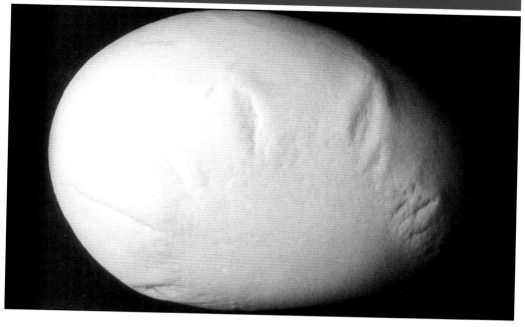

Description
Eggs are said to be flat-sided when part of the shell is flattened or indented. Often the adjoining part of the shell is wrinkled.

Incidence
Flat-sided eggs normally make up less than 1% of total production. They are most commonly produced by pullets in early lay, and may be the result of double ovulation or being held over an extra day in the shell gland. Incidence can vary with the strain of bird.

Cause	Control
Disease; traditionally linked with infectious bronchitis	Follow an effective vaccination programme.
Stress, e.g. frights and disturbances	To avoid frightening birds, minimise human activity in and around the shed. Increase shed security to stop other birds and animals entering the shed.
Crowding	Avoid overstocking.
Incorrect or changes in the lighting programme	There should not be sudden, large increases in day length as pullets come into lay.

Body-checked eggs

Description
The shells of body-checked eggs are marked by grooves and ridges called 'checks'. These are usually at the ends of the egg, especially the pointed end.

Checks at the middle, or 'waist', often completely encircle the egg. Such defects are the result of repairs to damage caused by stress or pressure when the egg is in the shell gland. This stage of development usually occurs in the last hours of the light period or the first half of the dark period.

Incidence
The percentage of body-checked eggs increases with flock age: at 35 weeks it can be up to 1%; at 60 weeks it can be 9%.

Cause	Control
Ageing of bird	Keep flock age as low as economically possible.
Stress, e.g. frights and disturbances	Minimise human activity in and around the layer shed, particularly during the critical period of the last few hours of the light period and the first half of the dark period. Increase shed security to stop other birds and animals entering the shed.
Incorrect lighting programme	Avoid using lighting programmes that keep birds active during the critical period. Where possible, light periods should not be longer than 15 hours. Do not use lighting programmes which allow for a 'midnight snack'.
Crowding	Avoid overstocking, particularly as birds become older.
Disease	Make sure that layer stock come from parent stock vaccinated against infectious bursal disease.

Pimples

Description
Pimples are small lumps of calcified material on the egg shell. Some can be broken off easily without damage to the shell while others may leave a small hole in the shell.

Incidence
An incidence of about 1% of total production is common.

Cause	Control
Thought to be caused by foreign material in the oviduct, which may be associated with:	
Ageing of bird	Keep flock age as low as economically possible.
Poor nutrition	Prevent excess calcium intake in winter.
Strain of bird	Be aware that some strains may produce eggs with a higher incidence of this fault.

Pinholes

Description and incidence
Pinholes, or very small holes in the egg shell, usually affect less than 0.5% of total production.

Cause	Control
May result from faulty laying down of the egg shell or from pimples being knocked off the shell. Both problems are thought to be associated with:	
Ageing of bird	Keep flock age as low as economically possible.
Poor nutrition	Prevent excess calcium intake in winter.
Strain of bird	Choose a strain which produces eggs with a lower incidence of this fault.
Damage from toenail points or other small sharp projections	Make sure that cage floor slope is enough to allow eggs to roll away easily from the birds. Remove sharp projections on cages and roll-out trays.

Mottled or glassy shells

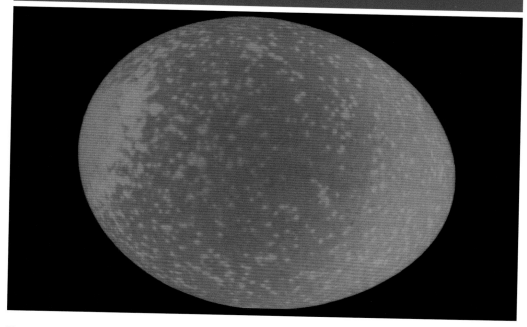

Description
When parts of the egg shell are translucent, it appears mottled or glassy. Such shells can also be thin and fragile.

The egg in the photograph has been placed over a bright light. This is the way it would look to a candling operator.

Incidence
The incidence is variable. Such eggs are not usually downgraded unless the condition is very obvious or the shells are thin.

Cause	Control
Failure of the egg shell to dry out quickly after laying, which is made worse by:	
High humidity in the layer shed	The shed must be well ventilated.
Crowding	Avoid overstocking.
Disease	Make sure that layer stock come from parent stock vaccinated against infectious bursal disease.

Cage marks

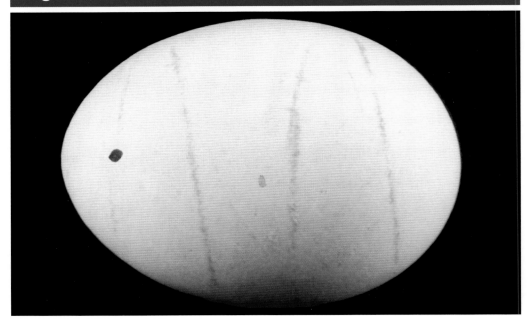

Description
'Cage marks' is the term used to refer to dirty marks, dirty lines or translucent lines on the shell when eggs are collected.

Incidence
In well-managed flocks, the incidence of this problem should be under 5%.

Cause	Control
Dirty marks or lines are due to:	
Rusty or dirty wires in the cage floor or roll-out trays	Check wire floors. If they are rusty, either: • replace the floors. • regalvanise if possible, or • paint with a light-coloured paint. Brush roll-out trays regularly to keep them clean. Do not allow excessive build-up of manure under cages. Remove manure caught in cage floors. Do not allow ammonia to build up in sheds, as it causes corrosion of metal cages. Make sure ventilation is adequate and that manure is kept dry.
Translucent lines result when the shell fails to dry out quickly after laying, which is made worse by:	
High humidity in the layer shed	The shed must be well ventilated.
Crowding	Avoid overstocking.

OPTIMUM EGG QUALITY - A PRACTICAL APPROACH

Stained eggs

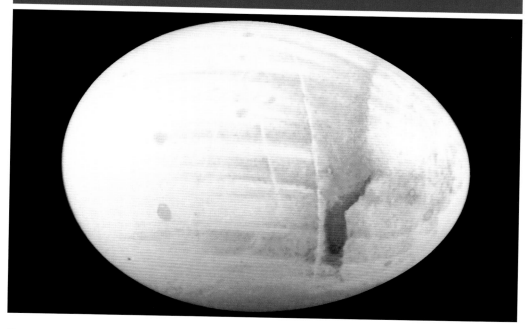

Description
All or part of the egg shell may become stained by various substances, e.g. blood, faeces.

Incidence
The incidence of stained eggs varies, partly because stains may be caused by a variety of substances. Smears of blood are more common on eggs from pullets in early lay.

Cause	Control
Blood from a prolapsed cloaca, cannibalism or vent picking	Do not allow pullets to become over-fat, as the incidence of prolapse is greater in fat birds. There should not be sudden large increases in day length as pullets come into lay. Regularly clean cage bottoms and roll-out trays. Clean belt pick-up systems.
Faecal contamination	Keep nest boxes supplied with clean nesting material. Maintain proper hygiene, follow effective vaccination programmes and, when necessary, use appropriate medication to keep birds free of diseases which cause enteritis. Feed should not contain high levels of ingredients causing loose or sticky droppings, such as molasses and high-tannin sorghums.

(continued)

Cause	Control
Water stains	Minimise roof condensation, which can drip onto eggs, by providing adequate shed ventilation. Eliminate drips from faulty misting nipples and drinker lines.
Sanitisers used in egg washing	Make sure that sanitisers in washing solutions are properly dissolved and used at the correct concentrations.
Grease and oil stains	Do not allow eggs to become contaminated by lubricants on rollers in packing systems.
Certain drugs produce mottled shells or white shells in breeds that normally lay brown eggs. The drug chlortetracycline produces yellow shells.	Do not feed to layers or pullets just before lay. Follow correct procedures for medication, including not feeding the medication for the required withholding time before eggs are collected, to prevent residues.

Fly marks

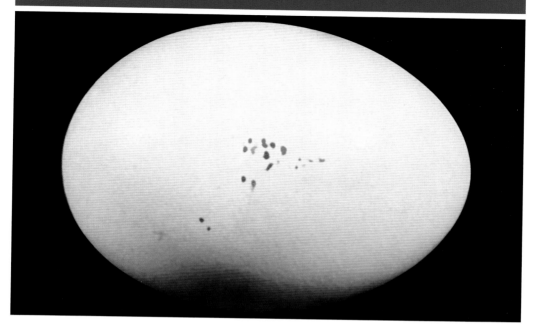

Description
Fly marks are caused by flies leaving droppings on the egg shell.

Incidence
This defect should not occur under good management. Any incidence is unacceptable.

Cause	Control
Eggs exposed to large numbers of flies	Control flies on the farm by maintaining a high level of hygiene, by keeping manure dry and not allowing it to build up excessively and, when necessary, by using approved insecticides. Use UV fly traps in egg-handling rooms. Put fly screens on doors and windows in handling, packing and storage rooms. Collect eggs frequently. Store eggs in a coolroom.

Fungus or mildew on shells

Description
Eggs affected by a fungus may have a green coating of powdery material or a black, beard-like growth on the shell. Sometimes such eggs are said to be affected by mildew.

Incidence
This defect should not occur under good management. Any incidence is unacceptable.

Cause	Control
Poor hygiene during handling, storage and transport	Do not use soiled egg-handling equipment. Regularly clean roll-out trays, collection belts and egg fillers. Use correct egg-washing sanitisers and procedures. Clean and disinfect coolrooms regularly.
Warm storage conditions or excess coolroom storage of eggs with other produce	The storage temperature must be under 15 °C, and the humidity in the storage room under 80%. Store only eggs in the coolroom.
Old eggs	Do not keep old eggs in handling and storage areas because they are more prone to fungal attack.

Internal defects

Blood spots

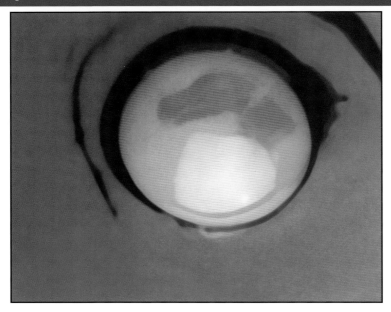

Description
Blood spots vary from barely distinguishable spots on the surface of the yolk to heavy blood contamination throughout the yolk. Occasionally blood may be diffused through the albumen or white of the egg.

Incidence
The incidence varies between strains of bird and can be as high as 10%. Between 2 and 4% of all eggs contain some blood.

Cause	Control
Blood vessels rupturing in the ovary or oviduct, affected by:	
Levels of vitamin A and vitamin K in the diet	Keep vitamin premixes cool and dry. If you prepare your own, they must be properly formulated and mixed.
Vitamin K antagonists (e.g. the drug sulphaquinoxaline and a component of lucerne meal)	Layer diets should not contain high levels of lucerne meal. Withdraw the drug sulphaquinoxaline from layers at least 10 days before collecting eggs for human consumption and follow any other requirements for correct medication.
Fungal toxins	Do not allow feed bins or feed lines to become contaminated by stale, wet or mouldy feed.
Lighting programme	Do not use continuous light, or light programmes consisting of short, intermittent light periods.
Frights and disturbances	Take care that birds are not disturbed by unusual or sudden loud sounds in the layer shed.
Avian encephalomyelitis	Follow an effective vaccination programme.
Strain of bird	Incidence of this fault may be higher in some strains.

Meat spots

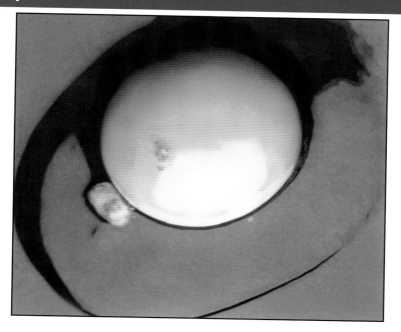

Description
Most meat spots are pieces of tissue from body organs, but some may be partially broken-down blood spots. They are usually brown in colour, and found in the thick albumen, chalazae, or the yolk. They range in size from 0.5 mm to more than 3 mm in diameter.

Incidence
The incidence of meat spots ranges from less than 3% to 30% or more. It varies with the strain of bird, increases with the age of bird and may be higher in brown eggs. Many meat spots are too small to be detected by candling, especially in brown eggs. Less than 1% of eggs are usually downgraded because of meat spots.

Cause	Control
Ageing of bird	Keep flock age as low as economically possible.
For meat spots derived from blood spots, see causes and controls of blood spots (page 48)	

Watery whites

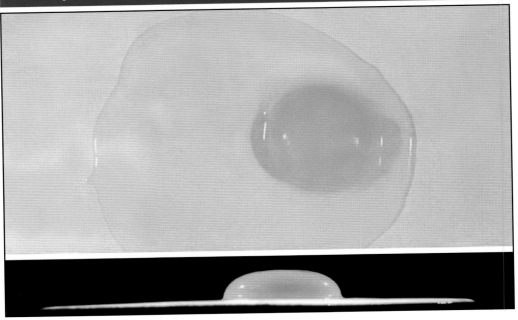

Description
When an egg broken onto a flat surface has a watery, spread-out white, this usually indicates that the egg is stale. The height of the white and the weight of the egg are used to calculate a value in Haugh units on a scale of 0 to 110; the lower the value, the staler the egg.

A minimum Haugh unit measurement of 60 is desirable for whole eggs sold to the domestic consumer. Most eggs leaving the farm should average between 75 and 85 Haugh units.

Incidence
The development of watery whites is chiefly due to the increasing age of the egg. The rate of development is increased by high storage temperature and low humidity (see figure 4). As birds age, the Haugh unit value of their eggs decreases by about 1.5 to 2 units per month of lay (see figure 3). Some birds consistently produce eggs with watery whites (Haugh units less than 30) later in lay.

Cause	Control
Old eggs	Minimise storage time on the farm by: • increasing the number of pick-ups per week for grading or delivery to retail outlets. • collecting and packing eggs laid on pick-up days so that they are consigned on that day. • grading eggs while they are fresh when packing them on the farm.

(continued)

Cause	Control
High storage temperature and low humidity	Reduce shed temperatures in summer. Collect eggs at least twice daily and even more frequently in summer. Store eggs in a coolroom at a temperature of less than 20 °C e.g. 12 to 15 °C (requirements may vary in different countries) as soon as possible after collection. If a humidifier is not fitted to the cooling unit, place an open tray of water in the coolroom to ensure humidity is kept at 70 to 80%. Oil eggs soon after collection. Use only an oil such as Caltex Pharma White 15 BP/USP approved for this purpose by the relevant authorities.
Ageing of bird	Keep flock age as low as economically possible.
Diseases, e.g. infectious bronchitis and egg drop syndrome	Follow effective vaccination programmes.
Fungal toxins	Do not allow feed bins or feed lines to become contaminated by stale, wet or mouldy feed.
Ammonia	Control ventilation to keep ammonia levels low.
Rough handling	Examine egg handling procedures and equipment, and modify to minimise bumping or shaking of eggs.
Incorrect packing	Pack eggs on filler flats with the air cell upwards.
Birds which persisently lay eggs with watery whites	Culling these birds is not practicable, as they are hard to identify. The only way of handling the problem is to remove the eggs at grading (like all eggs with watery whites, they are distinguished by an enlarged air cell).
Strain of bird	Be aware that some strains produce eggs with a high average Haugh unit rating.

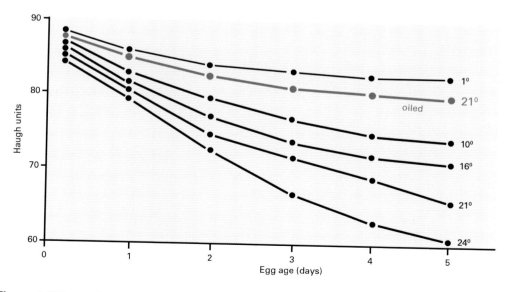

Figure 4: Effects of egg age and storage temperature on egg Haugh unit value

Pale yolks

Description
The colour of the yolk is due to substances called carotenoids. The nutritional value of the egg is not affected by the yolk colour. The intensity of yolk colour may be measured against standards such as the DSM Yolk Colour Fan. Most egg marketing authorities require deep-yellow to orange-yellow yolk colours in the range 9 to 12 on the DSM Yolk Colour Fan. Yolks of more intense colour may be required for specific markets.

The most important sources of carotenoids in poultry feed are maize (corn), maize gluten, alfalfa (lucerne) and grass meals; these sources contain the pigmenting carotenoids lutein and zeaxanthin, which, together with other oxygen-containing carotenoids, are known by the collective name of xanthophylls.

However, the carotenoid content in the ingredients of poultry feed is not constant; the pigmentation properties of the carotenoids can be weakened or lost in a variety of ways. These fluctuations in carotenoid content and availability concern both the poultry nutritionist and the feed producer. Because of such fluctuations, naturally-occurring carotenoids cannot be relied upon to provide the desired yolk colour or to provide a consistent colour. Therefore, nature-identical yellow and red carotenoids, such as apo-ester and canthaxanthin, are commonly added to feed in order to achieve the desired egg yolk colour. Consumed by the laying hen, these supplemental carotenoids are readily transferred to the blood and then deposited in the yolk to provide pigmentation.

Incidence
A wide variation in colour may normally be expected in the yolks from any flock. If a flock averages a yolk colour score of 10 on the DSM Yolk Colour Fan, two out of every three eggs laid by the flock will score between 9 and 11. Also, one egg in 20 will score less than 8, and one in 20 greater than 12.

Cause	Control
Deterioration of pigment concentrates	Keep concentrates and premixes in airtight containers, in a cool place and away from sunlight. Store for short periods only.
Insufficient carotenoids in the feed	The required yolk colour score, the percentage of egg production and daily feed consumption of the flock are the major factors governing the amount of pigment needed in the layer diet. Adjust levels of natural or synthetic carotenoids to meet these needs. Do not rely too much on feed ingredients such as lucerne or maize, whose pigment concentrations can vary widely.
Insufficient pigment in birds that are coming into lay	Flocks coming into lay may have insufficient stores of pigment. Start feeding pigment sources 2 to 3 weeks before the flock starts to lay.
Oxidising agents or pigment antagonists in the feed	The addition of a recommended antioxidant to feed will help preserve yolk pigments. Either do not include high levels of barley or triticale in the diet, or, if they are included, adjust pigment levels to higher than normal concentrations.
Inadequate mixing of feed	Always blend concentrates in a premix before adding to the rest of the feed for mixing. Make sure that mixing time is adequate.
Storage of feed in damp and/or hot conditions	Keep feed cool and dry.
Rough handling or transport of feed	Mixed feed should be handled carefully so that the various components do not separate out. This particularly needs to be checked when augers and automatic feeding systems are used.

Mottled yolks and discoloured yolks

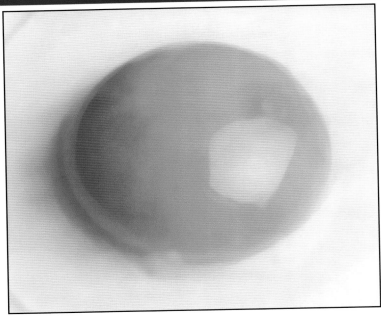

Description
The colour of the yolk may be uneven or patchy, as shown, or vary from the desired colour range (deep-yellow to orange-yellow).

Incidence
Some degree of mottling occurs in more than 50% of all eggs. The severity of mottling increases as eggs become stale. Discoloured yolks are rare.

Cause	Control
Mottled yolks have been attributed to:	
Poor handling and storage	Minimise storage time on the farm by: • increasing the number of pick-ups per week for grading or delivery to retail outlets. • collecting and packing eggs laid on pick-up days so that they are consigned on that day. • grading eggs while they are fresh when packing on the farm. Do not allow the temperature in the layer shed to become too high. Keep the coolroom temperature below 20 °C, e.g. at 12 to 15 °C (guidelines/requirements may vary in different countries) and the humidity at 70 to 80%. Examine egg handling procedures and equipment and modify to minimise rough handling. Oil eggs soon after collection. Use only an oil approved for this purpose by the relevant authorities.

(continued)

Cause	Control
Thin shells	Make sure that feed contains the correct amounts and proportions of nutrients, particularly calcium.
Gossypol in cottonseed meal	Limit cottonseed meal with high gossypol content to 5% of the layer diet (recent work indicates that up to 10% of some solvent-extracted cottonseed meal products can be safely used as they have low levels of gossypol due to newer varieties and improved processing). If the diet contains high gossypol cottonseed meal, add iron at the rate of 0.05% by weight of the diet. This can be achieved by adding ferrous sulphate powder at the rate of 0.25% by weight of the diet.
Treatment for worms	Do not use combinations of the worming drugs piperazine, phenothiazine and dibutyltin dilaurate immediately before or during lay. However, the use of any one of these drugs at the recommended and approved dose rate is not detrimental.
Certain antioxidants	Do not feed gallic acid or tannic acid to laying hens.
Raw soybean meal	Use only heat-treated soybean products in layer diets.
Discoloration of yolks has been attributed to:	
Gossypol in cottonseed meal	Prevent bluish-green yolks by limiting cottonseed meal with high gossypol content to 5% of the layer diet. New varieties and improved processing of CSM have been developed that have low levels of gossypol. If the diet contains cottonseed meal with high gossypol levels, add iron at the rate of 0.05% by weight of the diet. This can be achieved by adding ferrous sulphate powder at die rate of 0.25% by weight of the diet.
The weed shepherd's purse	Prevent green yolks by excluding this plant from the diet.
Cyclopropene fatty acids in cottonseed and certain weeds, e.g. marshmallow	These substances may cause the yolk to become a salmon colour after storage. Exclude weeds from the layer diet.
DPPD (diphenyl-para-phenylenediamine)	The antioxidant DPPD (diphenyl-para-phenylenediamine) (where approved for use) can cause excessive amounts of pigments from the feed to be deposited into the yolk.
Worms or disease	Disease or infestation with capillaria worms (cropworms) can produce 'blond' or 'platinum' yolks. Follow effective vaccination programmes against disease, observe hygiene measures and control worms.
Balance of yellow to red pigments	A low yellow:red carotenoid ratio in the feed, for example 1.25:1, will produce apricot-coloured yolks. Maintain a yellow:red ratio of approximately 3:1, particularly when diets are low in natural pigments.

Discoloured whites

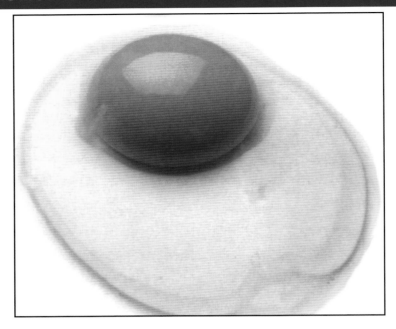

Description
The normal, slightly yellow-green colour of egg white may darken to an objectionable yellow or green, as shown, or may even become pink.

Incidence
This problem is rare.

Cause	Control
Excess of the vitamin riboflavin	Excess riboflavin in the diet will produce green whites. Give only the amount necessary to meet the birds' requirements.
Cyclopropene fatty acids in cottonseed and certain weeds, e.g. marshmallow	These substances may cause the white to become pink after storage. Limit cottonseed to 5% of the layer diet (solvent-extracted meals have low levels of cyclopropanoid fatty acid). Exclude weeds from the layer diet.
Ageing of the eggs and/or poor storage conditions	The whites of eggs stored for an extended period under poor conditions, e.g. at unsuitably high temperatures, may become much yellower. Make sure eggs are stored under optimal conditions, and minimise storage time.

Rotten eggs

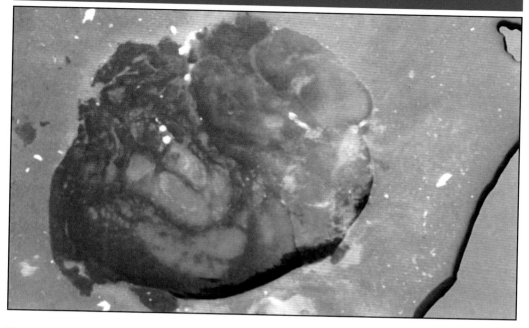

Description
Bacterial or fungal contamination of the egg can produce black, red or green rots. The egg looks and smells putrid when broken out.

Incidence
Under conditions of good management, the incidence of rotten eggs is very low, rising slightly in summer. Such eggs must be eliminated during grading, as they have such a detrimental effect on product image. It is unacceptable for even one rotten egg to reach the consumer.

Cause	Control
Faecal contamination	Follow relevant food safety handling and processing guidelines and requirements — details vary in different countries.
Improper washing procedures	Keep cage floors, collection belts and nest boxes free of droppings and egg contents. Brush roll-out trays frequently to keep them clean. Reject extremely dirty eggs. Wash dirty eggs only. The temperature of water used for washing should be 41—44 °C. Use only a recommended, registered egg-washing detergent-sanitiser combination at the correct concentration. For bubble-type washers, change the washing solution every 3 to 5 baskets; otherwise, change the solution every 4 hours. Dry and cool eggs immediately after washing. Thoroughly clean and sanitise washing equipment daily, using a different detergent-sanitiser mixture from that used to clean eggs. Do not wipe eggs with a damp cloth or rub them with abrasives.

(continued)

Cause	Control
High storage temperature and humidity	Reduce shed temperatures in summer. Collect eggs at least twice a day and even more often in summer.
Old eggs	As soon as possible after collection, store eggs in a coolroom at a temperature of 12 to 15 °C and a relative humidity of 70 to 80%. Make sure that eggs are not trapped in cages. Such eggs may eventually roll out and be collected when they are stale. If layers are kept on deep litter, collect eggs only from regularly used nests. Eggs discovered elsewhere may not be fresh. Reject any eggs whose freshness is in doubt.
Infection in a hen's oviduct	Such infections are rare. If they occur, make every effort to identify and cull the affected birds.

Roundworms in eggs

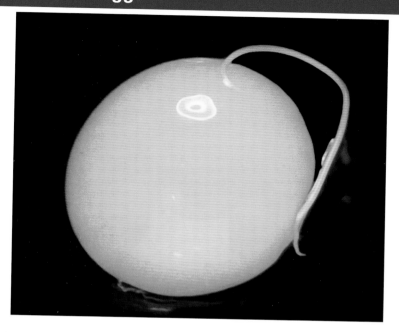

Description
Some eggs may contain one or more roundworms. These nematodes are intestinal parasites of hens.

Incidence
The incidence is quite rare.

Cause	Control
Roundworm infestation of the intestinal tract. Roundworms migrate from the cloaca into the oviduct, where they may be enclosed in the egg.	Protect the flock from serious worm infestation by: • feeding a well-balanced diet. • preventing fouling of feed and water. • keeping birds off fouled or damp ground or litter. • regularly and thoroughly cleaning and disinfecting the poultry house. • keeping chickens separate from adult birds. Rearing and housing birds on wire reduces worm infestation. Deworm pullets before caging and treat layers if necessary. Check cull birds for roundworms. This will give an idea of the incidence of infestation in the flock.

Off odours and flavours

Description
Some eggs may have an unusual or unacceptable odour or taste, although their appearance is normal. They differ from rotten eggs, which are obviously defective and smell putrid.

Incidence
Off odours and flavours are rare in fresh eggs stored correctly.

Cause	Control
Old eggs	Avoid the development of 'stale odours' by minimising storage time. High temperatures accelerate the development of 'stale odours'. Reduce temperatures in the layer shed in summer. Collect eggs at least twice daily and even more frequently in summer.
High storage temperatures	Store eggs in a coolroom as soon as possible after collection, preferably at a temperature of 12 to 15 °C. Oil eggs soon after collection. Use only an oil such as Caltex Pharma White 15 BP/USP, approved for this purpose by the relevant authorities.

(continued)

Cause	Control
Poor storage conditions	Absorption of foreign odours from the storage environment can produce off odours and tastes in the egg. Among materials implicated are: • fish oils and meals. • sour milk. • strongly scented or decaying fruit and vegetables. • mould. • disinfectants. • kerosene. • poultry droppings. Eggs that have been oiled are less likely to absorb foreign odours.
Strongly flavoured feed ingredients	'Fishy' or other undesirable flavours may be produced by feeding: • excessive amounts of low-grade fish meals or fish oils. • some vegetables including onions, turnips and excessive amounts of cabbage. • rapeseed.
Micro-organisms	Certain bacteria and fungi growing either on the outside or the inside of the egg may give an undesirable odour or flavour to the egg contents without causing noticeable spoilage. Use the same control measures as to minimise the incidence of rotten eggs (see page 57), e.g. avoid faecal contamination, use proper washing procedures, store eggs correctly and make sure they are as fresh as possible.
Persistent layers of off-flavoured eggs	Individual birds can continue to produce off-flavoured eggs regardless of feed and egg storage conditions. Such birds should be culled, but tracing them is extremely difficult.

Glossary

Air cell – The air space between the inner and outer shell membranes. Usually found at the broader end of the egg.

Albumen – The white of the egg.

Antagonist – An anti-nutritional factor present in feed which reduces its nutritional value.

Calcification – See Shell calcification.

Cannibalism – The vice of eating other birds.

Chalazae – White, twisted, rope-like structures attached to the thick albumen which anchor the yolk in the centre of the egg.

Cloaca – The common external opening for the digestive, urinary and reproductive tracts.

Cull – To remove unprofitable or otherwise unwanted birds from the flock.

Cuticle – The outer membrane covering the egg shell which gives the shell a bloom.

Filler – A moulded tray for packing eggs in an upright position.

Follicle – A sac-like structure in the ovary containing a developing ovum.

Germinal disc – A small, circular, white patch on the upper surface of the yolk, containing cells from which the chick develops if the egg is fertilised.

Grading – The process of classifying eggs by quality and by weight.

Infundibulum – A funnel-shaped section of the oviduct, about 6 to 8 cm long, which collects the ovum as it is released from the ovary.

Isthmus – A section of the oviduct about 12 cm long which secretes the shell membrane and some albumen.

Light programme – The schedule of artificial lighting for poultry flocks. Day length and light intensity are regulated to improve egg production.

Magnum – Part of the oviduct, immediately after the infundibulum. It is about 30 cm long and secretes about 40% of the egg's albumen.

Ovary – The female reproductive organ in which the ova (yolks) develop. Only the left ovary is functional in hens.

Oviduct – A long, tubular organ which adds the egg white, shell membranes and shell to the yolk as it passes through.

Ovulation – The release of a mature ovum from an ovarian follicle.

Ovum (plural ova) - An egg at the stage when it is a yolk without any white (albumen).

Oxidising agent – A substance which brings about chemical changes involving oxygen. In the case of poultry feed, these changes can result in loss of potency or nutritional value.

Premix – A blend of micro-ingredients, usually vitamins or trace minerals and including a carrier substance, added to feed during the mixing process.

Prolapse – The irreversible protrusion of the cloaca.

Pullet – A female bird in her first laying season.

Roll-out tray – An extension of the cage floor on which eggs accumulate for collection.

Shell calcification – The process of shell formation during the period the egg is in the shell gland (uterus).

Shell membranes – Inner and outer membranes that together line the inside of the shell and adhere very closely to it.

Uterus (shell gland) – Part of the oviduct, about 12 cm long, which secretes water into the albumen and forms the shell.

Vagina - Part of the oviduct, after the uterus, in which the shell cuticle is formed.

Vent (anus) – The external opening of the cloaca.

Vent picking – The vice of pecking other birds around the vent.

Vitelline membrane – A thin, transparent membrane which surrounds and contains the yolk.